SELECTED FROM

KON-TIKI

~

Thor Heyerdahl

Supplementary material by Ed Lavitt,
Peter Zahler and the staff of
Literacy Volunteers of New York City

OURWORLD™
Readers House
Literacy Volunteers of New York

OurWorld™ was made possible by grants from an anonymous foundation; Exxon Corporation; Scripps Howard Foundation; and H. W. Wilson Foundation.

ATTENTION READERS: We would like to hear what you think about our books. Please send your comments or suggestions to:
  The Editors
  Literacy Volunteers of New York City
  121 Avenue of the Americas
  New York, NY 10013

First LVNYC Printing: August 1993
ISBN: 1-56853-001-3

OurWorld is a series of books published by Readers House, the publishing division of Literacy Volunteers of New York City Inc., 121 Avenue of the Americas, New York, NY 10013. The word, "OurWorld" is a trademark of Readers House/Literacy Volunteers of New York City. READERS HOUSE and colophon are trademarks of Literacy Volunteers of New York City.

Cover designed by Kevin Barry; interior designed by Helene Berinsky.

Executive Director, LVNYC: Lilliam Barrios-Paoli
Publishing Director, LVNYC: Nancy McCord
Managing Editor: Sarah Kirshner
Publishing Coordinator: Yvette Martinez-Gonzalez
Marketing and Producting Manager: Elizabeth Bluemle

Our thanks to the LVNYC Board of Directors' Publishing Committee: James E. Galton, Geraldine E. Rhoads, Arnold Schaab, Martin Singerman and James Stanko.

LVNYC is an affiliate of Literacy Volunteers of America.

Acknowledgments

Literacy Volunteers of New York City gratefully acknowledges the generous support of the following foundations and corporations that made the publication of READERS HOUSE books possible: an anonymous foundation; Exxon Corporation; Scripps Howard Foundation; and H. W. Wilson Foundation.

This book could not have been realized without the kind and generous cooperation of the author, Thor Heyerdahl.

We deeply appreciate the contributions of the following suppliers: Creative Graphics (text typesetting); Delta Corrugated Container (corrugated display); Domtar Industries (text stock); Horizon Paper Company (cover stock); Offset Paperback Manufacturers, A Bertelsmann Company (cover and text printing and binding); and Phototype Color Graphics (cover color separations).

Thanks also to Joy M. Gannon and Claire Walsh of St. Martin's Press for producing this book, Marilyn Boutwell for her time devoted to searching for selections; Ed Lavitt and Peter Zahler for their skill and diligence in the research and writing of the supplementary material for this book; and to Marlene Charnizon for her thoughtful copyediting and suggestions. Thanks also to Barbara Mancuso, New York Times Pictures; Hawaii Tourist Bureau; and Aina Saetre of the Kon-Tiki Museum, Oslo, for help in obtaining photos.

Our thanks to Kevin Barry for his inspired design of the covers of these books. Thanks also to Helene Berinsky for her sensitive design of the interior of this book. Thanks also to Lloyd Birmingham for his accomplished design of maps and diagrams.

The raft, *Kon-Tiki*, ready to sail from Peru to Polynesia. Thor Heyerdahl and five crew members built the raft from balsa wood logs, in the style of ancient Peruvian rafts.

Contents

———∼———

Note to the Reader

The study of people, their origins and how they live is called anthropology. Anthropologists are the scientists who study people. They study people alive today and people who lived in ancient times.

This series of books is called *OurWorld* because each book in the series tells you something about the physical world around you. Every book features the work of a well-known author who writes about science. Each book also tells you about the special area of science the writer studies. Knowing more about science may help you understand the things that shape the world as we know it. When you're reading this book, you may find some answers to questions you have had about our world and why it is the way it is. We hope you will want to explore the many aspects that make up our world—past, present and future.

This book has different chapters. The Contents page lists these chapters and the pages where they start. Thor Heyerdahl's writing is in the chapter, "Selected from *Kon-Tiki*," starting on page 21.

Reading the other chapters in this book can help you understand Thor Heyerdahl's writing better, especially some of the scientific ideas that he is writing about. You may want to read some of these chapters before or after reading the selections. Here is what these chapters contain:

• The chapter, "About the Selections from *Kon-Tiki*," on page 10, gives you a preview of what you will read in Thor Heyerdahl's writing from *Kon-Tiki*.

• In the selections from *Kon-Tiki*, there may be scientific words or words about boats or the ocean that are unfamiliar. In the Glossary on page 50, right after the selections, you can find the meaning of the words that appear in **bold** type in the selections. You may want to review the words in the Glossary to get familiar with them before you read the selections. Special terms about boats are in **bold** in the selections and you can find them on the diagram of the raft *Kon-Tiki* on page 19.

• The selections from *Kon-Tiki* are about anthropology. Thor Heyerdahl set off on his voyage across the Pacific to try to prove a theory about the migration of ancient people. The chapter, "About Anthropology," on page 52, gives you more information about this science.

• The chapter, "About Thor Heyerdahl," on page 56, tells you about Thor Heyerdahl's life.

Sometimes knowing about a writer's life helps you understand his writing better.

• The chapter, "Questions and Activities for the Reader," on page 59, helps you build on what you have read. It has some ideas for group discussion and activities if you want to learn more about anthropology. It also lists "Resources" on page 63, such as other books by Thor Heyerdahl and magazines and videos on anthropology.

• This book also has a map of the voyage of the *Kon-Tiki* on page 20.

Reading about science will help you to be an *active* reader. Here are some things you can do.

Before reading, read the back cover of the book, think about its title and look at the picture on the front. Take a moment to think about why you want to read the book and what you already know about anthropology.

While reading, you will probably come across a word that is difficult to understand. You can find meanings for many of the words about boats and the ocean used in the selections in the Glossary on page 50. Reviewing these words may help you understand the selections better. Think about what you are reading and ask yourself questions such as: Does this information support what I already know about ancient peoples? or How can I use this information to answer questions I have about ancient peoples?

After reading the selections and some of the other chapters, try some of the questions and activities in the chapter on page 59. They are meant to help you discover more about what you have read and how it relates to you.

The editors of *OurWorld* hope you will write to us. We want to know what you think about our books.

About the Selections from KON-TIKI

~

What would you do if you were a scientist with an interesting idea and no one believed you? Thor Heyerdahl was faced with just this problem. Heyerdahl is an anthropologist from Norway. In the 1930s he did research in Polynesia, a group of small islands in the Pacific Ocean. He was interested in the history of the people on these remote islands. The islands are in the middle of the ocean and far from any other land. Heyerdahl wanted to know how people had first come to settle these islands.

The Polynesian people told stories of an ancestor of all their people who was named Tiki. The legends said that Tiki came to the islands around 500 A.D. (about 2,400 years ago).

Heyerdahl also learned that "Tiki" was a hero in the stories of another ancient civilization. Long ago, a group of people in Peru, South America, were led by a king named Kon-Tiki. According to legends in Peru, Kon-

Tiki lost a war around 500 A.D. and escaped across the Pacific Ocean. Heyerdahl came to believe that the Polynesians' Tiki and the Peruvians' Kon-Tiki were the same person. Heyerdahl thought that somehow Kon-Tiki had crossed the 4,000 miles of open ocean from South America to the islands of Polynesia.

Heyerdahl had other reasons to think ancient people may have come from South America to Polynesia. Giant stone sculptures on some of the Polynesian islands looked like statues in South America. Some of the plants on the islands were like those in South America, too.

However, other scientists did not believe Heyerdahl's idea. They did not think that people could have come from South America to Polynesia. During the time of Kon-Tiki, the only boats the Peruvians had were rafts. No one thought that a person could cross the Pacific Ocean in a raft. Heyerdahl became convinced that there was only one way anybody would believe his idea. He must cross the Pacific Ocean from Peru to Polynesia on a raft.

Although most people thought this idea was crazy, Heyerdahl found five adventurers who were willing to go with him. They decided to build a raft exactly like the ancient Peruvian rafts. In 1947 they traveled to the South American forests to find materials. They went into the jungle and cut down giant balsa trees and floated the logs down a river to the coast of Peru. (They used fresh green logs rather than dried logs because that is what the ancient Peruvians used. They were glad of it later because the sap in the fresh logs protected the wood from the water.) Then they built the raft without the use of nails or metal. They used ropes, bamboo and vines instead. They named their raft *Kon-Tiki*, after the South American king.

When the raft was finished, almost everyone who saw it thought it would sink or tip over, and that the sailors would drown. Nobody thought the six men would get very far.

One thing in Heyerdahl's favor was the Humboldt Current. This is a giant river of water in the Pacific Ocean. The current travels in a huge circle from South America across the Pacific Ocean almost to Asia, and then back again. Heyerdahl hoped the current would push his raft across the ocean, much as a raft would travel down a river. He also thought the trade winds that blow from east to west would help the raft sail toward Polynesia.

As the selection begins, Heyerdahl describes

how they built the raft. The crew was ready to begin the voyage.

Perhaps reading the selections from *Kon-Tiki* will remind you of what it takes to prove something you think is true. Perhaps they will make you think how the people in different parts of the world might be connected and how that could have happened.

The crew of the *Kon-Tiki*. From left to right: Knut Haugland, Bengt Danielsson, Thor Heyerdahl, Erik Hesselberg, Torstein Raaby, Herman Watzinger.

Cast of Characters

Thor Heyerdahl wanted five other crew members for his trip across the Pacific Ocean. He felt six people could get along well on a small raft and they could share the job of steering it. Each could steer two hours by day and two hours by night. Here are the crew members in the order that they joined the expedition.

THOR HEYERDAHL

Heyerdahl is a Norwegian anthropologist. In 1937, he first became interested in how people had reached the Polynesian Islands. To prove his idea that people came from South America to Polynesia, he decided to sail across the Pacific Ocean in a raft in 1947. He had served in the army during World War II and, because of the war, had waited ten years to try his idea out. He was 33 years old. On the raft trip, he kept the logbook of each day's events and photographed and filmed the journey.

HERMAN WATZINGER

Watzinger was an engineer from Norway who met Heyerdahl in New York when he was planning the expedition. Because he had experience with technical instruments, Watzinger became the raft's technical chief. He recorded weather and ocean water observations. This was important, for few ships had ever traveled along the route the *Kon-Tiki* took.

KNUT HAUGLAND

Haugland was a Norwegian friend of Heyerdahl's. They had met during World War II. Haugland had been a radio operator during the war. Since he was a radio expert, he became one of the two raft radio operators. He radioed new weather data and other information gathered on the trip to far-off science stations. The radio was also important if the raft needed help from the outside world.

TORSTEIN RAABY

Raaby was another Norwegian friend of Heyerdahl's from World War II. He was also a radio operator. Raaby's adventurous spirit and radio experience were useful on the raft. Heyerdahl describes Raaby as a "cheery fellow with blue eyes and bristly fair hair."

ERIK HESSELBERG

Hesselberg was the only real sailor in the crew. He was a boyhood friend of Heyerdahl's from Norway. Hesselberg had gone to navigation school and had sailed around the world. He became the raft's navigator. He used measuring instruments to plot where the raft was at all times. He also chose the direction in which the raft was to sail. He was a good guitarist, and his music was enjoyed by the other crew members. Heyerdahl describes Hesselberg as "a big hefty chap and . . . full of fun."

BENGT DANIELSSON

Red-bearded Danielsson was the last crew member chosen for the trip. He met Heyerdahl while the raft was being built in Peru. Danielsson was a scientist from Sweden. He was an anthropologist like Heyerdahl. He was also interested in how ancient people moved from place to place. He was the only crew member who could speak Spanish. This proved useful during the building of the raft. Danielsson packed 73 books to take on the raft trip. During the trip Danielsson was the steward, responsible for storing and preparing the food, although cooking chores were shared by all the men.

CARGO

Some of the supplies that the crew brought aboard the raft were: 56 small water cans with 275 gallons of drinking water; large wicker baskets full of fruit, sweet potatoes and coconuts; a battery-operated shortwave radio; camera and film; one box for each crew member; military rations for six men for four months; a rubber raft; a parrot; a paraffin lamp; cooking utensils; a primus stove with fuel; a harpoon; sleeping bags; maps; ropes and material for repairing sails.

DIAGRAM OF THE RAFT KON-TIKI

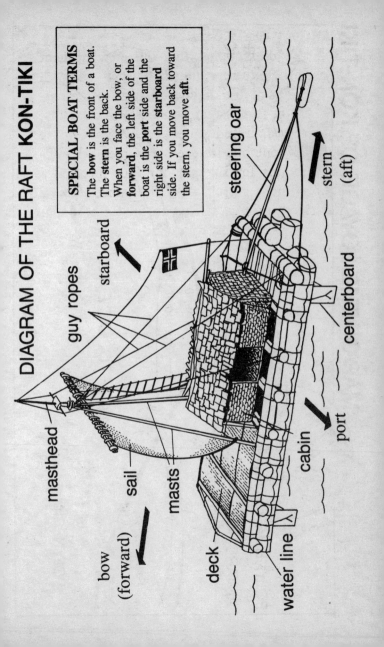

SPECIAL BOAT TERMS

The bow is the front of a boat. The stern is the back. When you face the bow, or forward, the left side of the boat is the port side and the right side is the starboard side. If you move back toward the stern, you move aft.

masthead

guy ropes

starboard

sail

masts

deck

water line

cabin

port

centerboard

steering oar

stern (aft)

bow (forward)

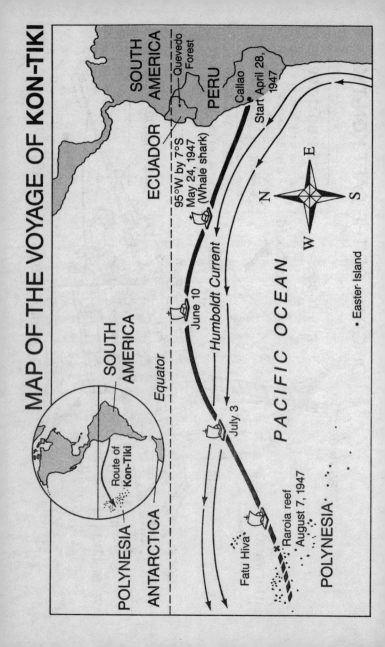

KON-TIKI

―――――― ~ ――――――

Thor Heyerdahl

For the first time for hundreds of years a **balsa** raft was being built in Callao Bay.

Nine of the thickest logs were chosen as sufficient to form the actual raft. Deep grooves were cut in the wood to prevent the ropes which were to fasten them and the whole raft together from slipping. Not a single spike, rail, or wire rope was used in the whole construction. The nine great logs were first laid loose side by side in the water so that they might all fall freely into their natural floating position before they were lashed securely together. The longest log, 45 feet long, was laid in the middle and projected a long way at both ends. Shorter and shorter logs were laid symmetrically on both sides of this, so that the sides of the raft were 30 feet long and the **bow** stuck out like a blunt plow.

The raft itself was now complete, laboriously fastened together with about three hundred different lengths of rope, each firmly knotted. A **deck** of split bamboos was laid upon it, fastened to it in the form of separate strips and covered with loose mats of plaited bamboo reeds. In the middle of the raft, but nearer the **stern**, we erected a small open **cabin** of bamboo canes, with walls of plaited bamboo reeds and a roof of bamboo slats with leathery banana leaves overlapping one another like tiles. **Forward** of the cabin we set up two **masts** side by side. They were cut from mangrove wood, as hard as iron, and leaned toward each other, so that they were lashed together crosswise at the top. The big rectangular square **sail** was hauled up on a **yard** made of two bamboo stems bound together to secure double strength.

At various places, where there were large chinks between the logs, we pushed down in all five solid fir planks which stood on their edges in the water under the raft.

The whole construction was a faithful copy of the old vessels in Peru and Ecuador.

All in all, there was little encouragement to be had from the experts who looked at the raft. Gales and perhaps hurricanes would wash us overboard and destroy the low, open craft, which would simply lie helpless and drift in circles about the ocean before wind and sea. Even in an ordinary choppy sea we should be continually drenched with salt water which would take the skin off our legs and ruin everything on board. If we added up all that the different experts, each in turn, had pointed out as the vital flaw in the construction itself, there was not a length of rope, not a knot, not a measurement, not a piece of wood in the whole raft which would not cause us to **founder** at sea.

These were difficult arguments to stifle. If only one of them proved to be right, we had not a chance. I am afraid I asked myself many times if we knew what we were doing. I could not counter the warnings one by one myself because I was not a seaman. But I had in reserve one single trump in my hand, on which the whole voyage was founded. I knew all the time in my heart that a prehistoric civilization had been spread from Peru and across to the islands at a time when rafts like ours were the only vessels on that coast. And I drew the general conclusion that, if balsa wood had floated and lashings held for Kon-Tiki in 500 A.D., they would do the same for us now if we blindly made our raft an exact copy of his.

A few days before the voyage was to start, the crew loaded their supplies onto the raft. On April 28, 1947, the raft Kon-Tiki *was towed 50 miles out to sea by the tugboat,* Guardian Rios, *to the Humboldt Current which was to carry them west.*

We were in the Humboldt Current, which carries its cold masses of water up from the Antarctic and sweeps them north all along the coast of Peru till they swing west and out across the sea just below the **Equator**.

The towrope was **cast off** and the raft was alone again. Thirty-five men on board the *Guardian Rios* stood at the rail waving for as long as we could distinguish outlines. And six men sat on the boxes on board the *Kon-Tiki* and followed the tug with their eyes as long as they could see her. Not till the black column of smoke had dissolved and vanished over the horizon did we shake our heads and look at one another.

"Good-by, good-by," said Torstein. "Now we'll have to start the engine, boys!"

We considered the possibility of paddling but agreed to wait for a wind.

And the wind came. It blew up from the southeast quietly and steadily. Soon the sail filled and bent forward like a swelling breast. And the *Kon-Tiki* began to move. We shouted westward ho! and hauled on **sheets** and ropes. The **steering oar** was put into the water, and the **watch roster** began to operate. We threw balls of paper and chips of wood overboard at the bow and stood **aft** with our watches.

"One, two, three . . . eighteen, nineteen—now!"

Paper and chips passed the steering oar and soon lay like pearls on a thread, dipping up and down in the **trough** of the waves astern. We went forward yard by yard. The *Kon-Tiki* did not plow through the sea like a sharp-prowed racing **craft**. Blunt and broad, heavy and solid, she splashed sedately forward over the waves. She did not hurry, but when she had once got going she pushed ahead with un-shakable energy.

By the late afternoon the **trade wind** was already blowing at full strength. It quickly stirred up the ocean into roaring seas which swept against us from astern. For the first time we fully realized that here was the sea itself come to meet us; it was bitter earnest now—our communications were cut. Whether things went well now would depend entirely on the balsa raft's good qualities in the open sea. We knew that, from now onward, we should never get another onshore wind or chance of turning back. We were in the path of the real trade wind, and every day would carry us farther and farther out to sea. The only thing to do was to go ahead under full sail; if we tried to turn homeward, we should only drift farther out to sea stern first. There was only one possible course, to sail **before the wind** with our bow toward the sunset. And, after all, that was the object of our voyage—to follow the sun in its path as we thought Kon-Tiki and the old sun-worshipers must have done when they were driven out to sea from Peru.

About midnight a ship's light passed in a northerly direction. At three another passed on the same **course**. We waved our little paraffin lamp and hailed them with flashes from an electric torch, but they did not see us and the lights passed slowly northward into the darkness and disappeared. Little did those on board realize that a real Inca raft lay close to them, tumbling among the waves. And just as little did we on board the raft realize that this was our last ship and the last trace of men we should see till we had reached the other side of the ocean.

The ocean currents and the winds increased. As the sea grew rougher, the crew took turns handling the sails and steering. After two-and-a-half days of round-the-clock steering and sailing, the sea grew calmer. The crew crept into the cabin and slept. The raft continued on its way.

We were now so accustomed to having the sea dancing round us that we took no account of it. What did it matter if we danced round a bit with a thousand **fathoms** of water under us, so long as we and the raft were always on top? It was only that here the next question arose—how long could we count on keeping on top? It was easy to see that the balsa logs absorbed water. The aft crossbeam was worse than the others; on it we could press our whole finger tip into the soaked wood till the water squelched. Without saying anything I broke off a piece of the sodden wood and threw it overboard. It sank quietly beneath the surface and slowly vanished down into the depths. Later I saw two or three of the other fellows do exactly the same when they thought no one was looking. They stood looking reverently at the waterlogged piece of wood sinking quietly into the green water.

We had noted the **water line** on the raft
when we started, but in the rough sea it was
impossible to see how deep we lay, for one moment the logs were lifted out of the water and
the next they went deep down into it. But, if
we drove a knife into the timber, we saw to our
joy that the wood was dry an inch or so below
the surface. We calculated that, if the water
continued to force its way in at the same pace,
the raft would be lying and floating just under
the surface of the water by the time we could
expect to be approaching land. But we hoped
that the sap further in would act as an impregnation and check the absorption.

Then there was another menace which trou-
bled our minds a little during the first weeks.
The ropes. In the daytime we were so busy that
we thought little about it, but, when darkness
had fallen and we had crept into bed on the
cabin floor, we had more time to think, feel,
and listen. As we lay there, each man on his
straw mattress, we could feel the reed matting
under us heaving in time with the wooden logs.
In addition to the movements of the raft itself
all nine logs moved reciprocally. When one
came up, another went down with a gentle
heaving movement. They did not move much,
but it was enough to make one feel as if one

were lying on the back of a large breathing animal, and we preferred to lie on a log lengthways. The first two nights were the worst, but then we were too tired to bother about it. Later the ropes swelled a little in the water and kept the nine logs quieter.

But all the same there was never a flat surface on board which kept quite still in relation to its surroundings. As the foundation moved up and down and round at every joint, everything else moved with it. The bamboo deck, the double mast, the four plaited walls of the cabin, and the roof of slats with the leaves on it—all were made fast just with ropes and twisted about and lifted themselves in opposite directions. It was almost unnoticeable but it was evident enough. If one corner went up, the other corner came down, and if one half of the roof dragged all its laths forward, the other half dragged its laths astern. And, if we looked out through the open wall, there was still more life and movement, for there the sky moved quietly round in a circle while the sea leaped high toward it.

The ropes took the whole pressure. All night we could hear them creaking and groaning, chafing and squeaking. It was like one single complaining chorus round us in the dark, each rope having its own note according to its thickness and tautness.

Every morning we made a thorough inspection of the ropes. We were even let down with our heads in the water over the edge of the raft, while two men held us tight by the ankles, to see if the ropes on the bottom of the raft were all right. But the ropes held. A fortnight the seamen had said. Then all the ropes would be worn out. But, in spite of this consensus of opinion, we had not so far found the smallest sign of wear. Not till we were far out to sea did we find the solution. The balsa wood was so soft that the ropes wore their way slowly into the wood and were protected, instead of the logs wearing the ropes.

After a week or so the sea grew calmer, and we noticed that it became blue instead of green. We began to go west-northwest instead of due northwest and took this as the first faint sign that we had got out of the coastal current and had some hope of being carried out to sea.

The very first day we were left alone on the sea we had noticed fish round the raft, but we were too much occupied with the steering to think of fishing. The second day we went right into a thick **shoal** of sardines, and soon afterward an eight-foot blue shark came along and rolled over with its white belly uppermost as it rubbed against the raft's stern, where Herman and Bengt stood barelegged in the seas, steering. It played round us for a while but disappeared when we got the hand harpoon ready for action.

Next day we were visited by **tunnies**, bonitos, and dolphins, and when a big flying fish thudded on board we used it as bait and at once pulled in two large dolphins (**dorados**) weighing from twenty to thirty-five pounds each. This was food for several days. On steering watch we could see many fish we did not even know, and one day we came into a school of porpoises which seemed quite endless. The black backs tumbled about, packed close together, right in to the side of the raft, and sprang up here and there all over the sea as far as we could see from the **masthead**. And the nearer we came to the Equator, and the farther from the coast, the commoner flying fish became. When at last we came out into the blue water where the sea rolled by majestically, sunlit and serene, ruffled by gusts of wind, we could see them glittering like a rain of projectiles which shot from the water and flew in a straight line till their power of flight was exhausted and they vanished beneath the surface.

If we set the little paraffin lamp out at night, flying fish were attracted by the light and, large and small, shot over the raft. They often struck the bamboo cabin or the sail and tumbled helpless on the deck. Unable to get a take-off by swimming through the water, they just remained lying and kicking helplessly, like large-eyed herrings with long breast fins. It sometimes happened that we heard an outburst of strong language from a man on deck when a cold flying fish came unexpectedly, at a good speed, slap into his face. They always came at a good pace and snout first, and if they caught one full in the face they made it burn and tingle. But the unprovoked attack was quickly forgiven by the injured party, for, with all its drawbacks, we were in a maritime land of enchantment where delicious fish dishes came hurling through the air. We used to fry them for breakfast.

The cook's first duty, when he got up in the forming, was to go out on deck and collect all the flying fish that had landed on board in the course of the night. There were usually half a dozen or more, and once we found twenty-six fat flying fish on the raft. Knut was much upset one morning because, when he was standing operating with the frying pan, a flying fish

struck him on the hand instead of landing right
in the cooking fat.

Watzinger shows off a bonito. The crew ate fish they
caught and fish that washed up on the raft.

The sea contains many surprises for him
who has his floor on a level with the surface
and drifts along slowly and noiselessly. A
sportsman who breaks his way through the
woods may come back and say that no wild life
is to be seen. Another may sit down on a
stump and wait, and often rustlings and crack-

lings will begin and curious eyes peer out. So it
is on the sea, too. We usually plow across it
with roaring engines and piston strokes, with
the water foaming round our bow. Then we
come back and say that there is nothing to see
far out on the ocean.

Not a day passed but we, as we sat floating
on the surface of the sea, were visited by in-
quisitive guests which wriggled and waggled
about us, and a few of them, such as dolphins
and pilot fish, grew so familiar that they ac-
companied the raft across the sea and kept
round us day and night.

It was May 24, and we were lying drifting on
a leisurely swell in exactly 95° west by 7°
south. I heard a wild war whoop from Knut,
who was sitting aft behind the bamboo cabin.
He bellowed "Shark!" till his voice cracked in a
falsetto, and, as we had sharks swimming
alongside the raft almost daily without creating
such excitement, we all realized that this must
be something extra-special and flocked astern
to Knut's assistance.

Knut had been squatting there, washing his
pants in the swell, and when he looked up for
a moment he was staring straight into the big-
gest and ugliest face any of us had ever seen in
the whole of our lives. It was the head of a ver-

itable sea monster, so huge and so hideous
that, if the Old Man of the Sea himself had
come up, he could not have made such an im-
pression on us. The head was broad and flat
like a frog's, with two small eyes right at the
sides, and a toadlike jaw which was four or five
feet wide and had long fringes drooping from
the corners of the mouth. Behind the head was
an enormous body ending in a long thin tail
with a pointed tail fin which stood straight up
and showed that this sea monster was not any
kind of whale. The body looked brownish
under the water, but both head and body were
thickly covered with small white spots.

The monster came quietly, lazily swimming
after us from astern. It grinned like a bulldog
and lashed gently with its tail. The large round
dorsal fin projected clear of the water and
sometimes the tail fin as well, and, when the
creature was in the trough of the swell, the water flowed about the broad back as though
washing round a submerged reef. In front of
the broad jaws swam a whole crowd of zebra-
striped pilot fish in fan formation, and large
remora fish and other parasites sat firmly attached to the huge body and traveled with it
through the water, so that the whole thing
looked like a curious zoological collection

crowded round something that resembled a floating deep-water reef.

When the giant came close up to the raft, it rubbed its back against the heavy steering oar, which was just lifted up out of the water, and now we had ample opportunity of studying the monster at the closest quarters—at such close quarters that I thought we had all gone mad, for we roared stupidly with laughter and shouted overexcitedly at the completely fantastic sight we saw. Walt Disney himself, with all his powers of imagination, could not have created a more hair-raising sea monster than that which thus suddenly lay with its terrific jaws along the raft's side.

The monster was a whale shark, the largest shark and the largest fish known in the world today. It is exceedingly rare, but scattered specimens are observed here and there in the tropical oceans. The whale shark has an average length of fifty feet, and according to zoologists it weighs fifteen tons. It is said that large specimens can attain a length of sixty-five feet; one harpooned baby had a liver weighing six hundred pounds and a collection of three thousand teeth in each of its broad jaws.

Our monster was so large that, when it began to swim in circles round us and under the raft, its head was visible on one side while the whole of its tail stuck out on the other. And so incredibly grotesque, inert, and stupid did it appear when seen fullface that we could not help shouting with laughter, although we realized that it had strength enough in its tail to smash both balsa logs and ropes to pieces if it attacked us. Again and again it described narrower and narrower circles just under the raft, while all we could do was to wait and see what might happen. When it appeared on the other side, it glided amiably under the steering oar and lifted it up in the air, while the oar blade slid along the creature's back.

In reality the whale shark went on encircling us for barely an hour, but to us the visit seemed to last a whole day. At last it became too exciting for Erik, who was standing at a corner of the raft with an eight-foot hand harpoon, and, encouraged by ill-considered shouts, he raised the harpoon above his head. As the whale shark came gliding slowly toward him and its broad head moved right under the corner of the raft, Erik thrust the harpoon with all his giant strength down between his legs and deep into the whale shark's gristly head. It was a second or two before the giant understood properly what was happening. Then in a flash the placid half-wit was transformed into a mountain of steel muscles.

We heard a swishing noise as the harpoon line rushed over the edge of the raft and saw a cascade of water as the giant stood on its head and plunged down into the depths. The three men who were standing nearest were flung about the place, head over heels, and two of them were flayed and burned by the line as it rushed through the air. The thick line, strong enough to hold a boat, was caught up on the side of the raft but snapped at once like a piece of twine, and a few seconds later a broken-off harpoon shaft came up to the surface two hundred yards away. A shoal of frightened pilot fish shot off through the water in a desperate attempt to keep up with their old lord and master. We waited a long time for the monster to come racing back like an infuriated submarine, but we never saw anything more of him.

Weeks passed. No ships or drifting objects were seen. The raft was on an empty sea. Each crew member did his job: maintaining radio contact with the outside world, taking care of the raft, steering, cooking and plotting the raft's location each day.

On June 10, the raft was closest to the Equa-

tor. On the 45th day at sea, Erik, the navigator, reported that they were halfway across the Pacific. The raft was 2,000 miles from its starting place in Peru to the east, and the first islands were 2,000 miles to the west.

The sea and wind remained unchanged, and the raft made smooth progress for many days. At the beginning of July, the first storm hit the raft, but the raft and crew weathered it well. Rain allowed the crew to collect fresh drinking water and to wash some of the salt off their bodies.

Two weeks later, the wind became very strong and Herman fell into the water and could not catch up with the raft as it moved away from him. He was rescued by Knut. For five days the Kon-Tiki was battered by a wild storm and a gale. On the fifth day, the sun broke through the clouds. The men and their cargo were safe.

After the two storms the Kon-Tiki had become a good deal weaker in the joints. The strain of working over the steep wave backs had stretched all the ropes, and the continuously working logs had made the ropes eat into the balsa wood. We thanked Providence that we had followed the Incas' custom and had not used wire ropes, which would simply have sawed the whole raft into matchwood in the gale. And, if we had used bone-dry, high-floating balsa at the start, the raft would long ago have sunk into the sea under us, saturated with sea water. It was the sap in the fresh logs which served as an impregnation and prevented the water from filtering in through the porous balsa wood.

But now the ropes had become so loose that it was dangerous to let one's foot slip down between two logs, for it could be crushed when they came together violently. Forward and aft, where there was no bamboo deck, we had to give at the knees when we stood with our feet wide apart on two logs at the same time. The logs aft were as slippery as banana leaves with wet seaweed, and, even though we had made a regular path through the greenery where we usually walked and had laid down a broad plank for the steering watch to stand on, it was not easy to keep one's foothold when a sea struck the raft. On the **port** side one of the nine giants bumped and banged against the cross-beams with dull, wet thuds both by night and by day. There came also new and fearful creakings from the ropes which held the two sloping masts together at the masthead, for the steps of the masts worked about independently of each other, because they rested on two different logs.

It was useless to try to inspect the ropes on the underside, for they were completely overgrown with seaweed. On taking up the whole bamboo deck we found only three of the main ropes broken; they had been lying crooked and pressed against the cargo, which had worn them away. It was evident that the logs had absorbed a great weight of water but, since the cargo had been lightened, this was roughly canceled out. Most of our provisions and drinking water were already used up, likewise the radio operators' dry batteries.

Nevertheless, after the last storm it was clear enough that we should both float and hold together for the short distance that separated us from the islands ahead. Now quite another problem came into the foreground—how would the voyage end?

The *Kon-Tiki* would slog on inexorably westward until she ran her bow into a solid rock or some other fixed object which would stop her drifting. But our voyage would not be ended until all hands had landed safe and sound on one of the numerous Polynesian islands ahead.

The crew now had to worry about what lay ahead. The raft was headed toward two groups of islands. One was 300 miles to the northwest.

The other group was 300 miles to the southwest. Depending on the wind and current, the raft could land on these islands or miss them entirely. Both groups of islands presented problems. One group of islands had steep cliffs at the water's edge, and the other had dangerous reefs.

Currents were starting to shift. In the next weeks, large flocks of birds circled around the raft. Birds meant land was nearby. At evening, just before sunset, the birds returned to land. The crew turned the raft in the direction the birds were flying.

On July 30, at daybreak, the crew saw land for the first time. They had reached Polynesia. Unfortunately, they couldn't land on the island they could see. The current had carried them away from the island during the night.

Four days later, the raft drew close to another island. The island was surrounded by a reef just under the surface of the water. If the raft landed on the reef, it could wreck the raft.

The crew zigzagged the raft alongside the island looking for an opening in the reef. A canoe with two people from the island in it shot through an opening in the reef and came toward the raft. The wind shifted and the raft drifted away from the island. The islanders in the canoe tried to tow the raft back to the *island, but the wind and the current pushed the raft away. The islanders returned to their village on the island.*

The raft continued to drift for three more days. No land was seen. The Kon-Tiki *was now headed for a dangerous 40- to 50-mile-long reef: Raroia reef. If the raft hit the reef, it would be wrecked.*

When night came, we had been a hundred days at sea.

Late in the night I woke, feeling restless and uneasy. There was something unusual in the movement of the waves. The *Kon-Tiki's* motion was a little different from what it usually was in such conditions. We had become sensitive to changes in the rhythm of the logs. I thought at once of suction from a coast, which was drawing near, and was continually out on deck and up the mast. Nothing but sea was visible. But I could get no quiet sleep. Time passed.

At dawn, just before six, Torstein came hurrying down from the masthead. He could see a whole line of small palm-clad islands far ahead. Before doing anything else we laid the oar over to southward as far as we could. What Torstein had seen must be the small **coral** islands which lay strewn like pearls on a string behind the Raroia reef. A northward current must have caught us.

We were drifting diagonally right in toward the reef. With fixed **centerboards** we should still have had some hope of steering clear. But sharks were following close astern, so that it was impossible to dive under the raft and tighten up the loose centerboards with fresh **guy ropes**.

We saw that we had now only a few hours more on board the *Kon-Tiki*. They must be used in preparation for our inevitable wreck on the coral reef. Every man learned what he had to do when the moment came; each one of us knew where his own limited sphere of responsibility lay, so that we should not fly round treading on one another's toes when the time came and seconds counted. The *Kon-Tiki* pitched up and down, up and down, as the wind forced us in. There was no doubt that here was the turmoil of waves created by the reef—some waves advancing while others were hurled back after beating vainly against the surrounding wall.

On board the *Kon-Tiki* all preparations for the end of the voyage were being made. Everything of value was carried into the cabin and lashed fast. Documents and papers were packed into watertight bags, along with films and other things which would not stand a dip in the sea. The whole bamboo cabin was covered with canvas, and especially strong ropes were lashed across it.

Order number one, which came first and last, was: Hold on to the raft! Whatever happened, we must hang on tight on board and let the nine great logs take the pressure from the reef. We ourselves had more than enough to do to withstand the weight of the water. If we jumped overboard, we should become helpless victims of the suction which would fling us in and out over the sharp corals. The rubber raft would capsize in the steep seas or, heavily loaded with us in it, it would be torn to ribbons against the reef. But the wooden logs would sooner or later be **cast ashore**, and we with them, if we only managed to hold fast.

Next, all hands were told to put on their shoes for the first time in a hundred days and to have their life belts ready. The last precaution, however, was not of much value, for if a man fell overboard he would be battered to death, not drowned. We had time, too, to put our passports and such few dollars as we had left into our pockets. But it was not lack of time that was troubling us.

When we realized that the **seas** had got hold of us, the **anchor** rope was cut and we were off. A sea rose straight up under us, and we felt the *Kon-Tiki* being lifted up in the air. The great moment had come; we were riding on the wave back at breathless speed, our ramshackle craft creaking and groaning as she quivered under us. The excitement made one's blood boil. I remember that, having no other inspiration, I waved my arm and bellowed "Hurrah!" at the top of my lungs; it afforded a certain relief and could do no harm anyway. The others certainly thought I had gone mad, but they all beamed and grinned enthusiastically. On we ran with the seas rushing in behind us; this was the *Kon-Tiki's* baptism of fire. All must and would go well.

But our elation was soon dampened. A new sea rose high up astern of us like a glittering, green glass wall. As we sank down it came rolling after us, and, in the same second in which I saw it high above me, I felt a violent blow and was submerged under floods of water. I felt the suction through my whole body, with such great power that I had to strain every single muscle in my frame and think of one thing only—hold on, hold on! I think that in such a desperate situation the arms will be torn off before the brain consents to let go, evident as the outcome is. Then I felt that the mountain of water was passing on and relaxing its devilish grip of my body. When the whole mountain had rushed on, with an ear-splitting roaring and crashing, I saw Knut again hanging on beside me, doubled up into a ball. Seen from behind, the great sea was almost flat and gray. As it rushed on, it swept over the ridge of the cabin roof which projected from the water, and there hung the three others, pressed against the cabin roof as the water passed over them.

We were still afloat.

Then I saw the next sea come towering up, higher than all the rest, and again I bellowed a warning aft to the others as I climbed up the **stay**, as high as I could get in a hurry, and hung on fast. Then I myself disappeared sideways into the midst of the green wall which towered high over us. The others, who were farther aft and saw me disappear first, estimated the height of the wall of water at twenty-five feet, while the foaming crest passed by fifteen feet above the part of the glassy wall into which I had vanished. Then the great wave reached them, and we had all one single thought—hold on, hold on, hold, hold, hold!

We must have hit the reef that time. I myself
felt only the strain on the stay, which seemed
to bend and slacken jerkily. But whether the
bumps came from above or below I could not
tell, hanging there. The whole submersion
lasted only seconds, but it demanded more en-
durance than we usually have in our bodies.
There is greater strength in the human mecha-
nism than that of the muscles alone. I determined that, if I was to die, I would die in this
position, like a knot on the stay. The sea thundered on, over and past, and as it roared by it
revealed a hideous sight. The *Kon-Tiki* was
wholly changed, as by the stroke of a magic
wand. The vessel we knew from weeks and
months at sea was no more; in a few seconds
our pleasant world had become a shattered
wreck.

Where we had stranded, we had only pools
of water and wet patches of coral about us; farther in lay the calm blue lagoon. The **tide** was
going out, and we continually saw more corals
sticking up out of the water round us, while the
surf which thundered without interruption
along the reef sank down, as it were, a floor
lower. What would happen there on the narrow
reef when the tide began to flow again was un-
certain. We must get away.

The reef stretched like a half-submerged fortress wall up to the north and down to the south. In the extreme south was a long island densely covered with tall palm forest. And just above us to the north, only 600 or 700 yards away, lay another but considerably smaller palm island. It lay inside the reef, with palm tops rising into the sky and snow-white sandy beaches running out into the still lagoon. The whole island looked like a bulging green basket of flowers, or a little bit of concentrated paradise.

This island we chose.

Herman stood beside me beaming all over his bearded face. He did not say a word, only stretched out his hand and laughed quietly. The *Kon-Tiki* still lay far out on the reef with the spray flying over her. She was a wreck, but an honorable wreck. Everything above deck was smashed up, but the nine balsa logs from the Quevedo forest in Ecuador were as intact as ever. They had saved our lives.

Glossary

—∿—

anchor. A heavy object attached to a boat or ship by a cable or rope. The anchor is thrown overboard to keep the boat in place.

balsa. A strong but lightweight wood found in tropical America.

before the wind. In the same direction as the wind.

cast ashore. Thrown onto land from the sea.

cast off. To untie a boat.

coral. A stone-like material which is formed from the bodies of sea animals. Reefs around islands in the Pacific are often formed by coral.

course. The path or direction in which something moves.

craft. A ship, a boat or a raft.

dorados. Large dolphins that live in warm salt water.

dorsal. Located on the back of an animal.

Equator. An imaginary line around the Earth halfway between the North and South Poles.

fathoms. Fathoms are a unit of measurement of the depth of water. One fathom equals six feet.

founder. To take on water and sink.

seas [sea]. Large bodies of salty water. "Seas" can also mean big waves and rough water.

sheets. The sails on a boat.

shoal. A group of fish.

stay. A rope or cable used as a support for a mast.

tide. The rise and fall of water along a seacoast. Tides change about every six hours. High tide is the highest point the sea comes up on the beach. Low tide is the lowest point the water moves out from the shore. Tides are caused by the pull of the Moon and the Sun on the Earth.

trade wind. Strong, steady winds over the ocean that blow toward the Equator.

trough. A low point between waves.

tunnies. Tuna fish.

watch roster. The list telling when each crew member had to help steer the raft.

yard. A long horizontal pole used to hold up a sail.

About Anthropology

People study many different things in order to understand the world around us. A person who studies people is called an anthropologist.

The study of human beings—ourselves—is important to us. There are many questions we want answered about ourselves. Where did we come from? How did people live hundreds or thousands of years ago? Why do we live in groups—families, tribes, communities? How do we differ from each other and how are we the same? Anthropologists try to answer questions such as these.

Why is it important to find answers to these questions? People want to know who their ancestors are and where they came from. And we want to know where humans as a group came from and why we are the way we are today.

Anthropologists study what different people look like and how they behave. They also study the things that human beings make and use, like tools and clothing. They study what human beings believe and value. Anthropologists have all of humankind to study—past and present.

Physical anthropology is the study of the physical structure of humans and how and why it has evolved. Physical anthropologists might examine ancient fossils (bones or traces of bones in rock) of primitive humans to see how they differ from the bones of modern humans. Or they might study the behavior of our near relatives, the apes and monkeys, to try and understand early and modern human behavior. Or they might research the diets of people from around the world and how different foods affect their health.

Archaeologists are anthropologists who study remains of ancient civilizations. They often try to figure out how people lived in the past from clues found in fragments of pots or ruins of buildings or even trash piles outside prehistoric caves. Archaeologists explain how our ancient ancestors lived.

Cultural anthropology is the study of existing human cultures. A culture is a group of people whose art, language, style of living or some other feature sets them apart from other groups of people. For example, Americans have a common culture; yet people who live in different areas of America can have different cultures, too. Cultural anthropologists are interested in comparing the languages, stories and customs of human beings in the present and the recent past.

A **linguist** is an anthropologist who mainly studies languages. Language can tell a lot about a culture. For example, if there are many words describing a certain animal, it may show that the animal plays an important role in the culture, perhaps as food or as an object of worship. Language can describe how a society is organized. Linguists are interested in how different languages are put together and how they developed.

An **ethnographer** is an anthropologist who studies everything about one living culture—its religion, marriage customs, social functions, and the like. Ethnographers often go and live among the people of the culture they want to study to observe the way the people live.

Ethnologists may use the studies of ethnographers to analyze different cultures and to compare them. They study the origins of different peoples and what distinguishes one from another. Thor Heyerdahl studied the stories and art of the culture for clues about where the people of the South Pacific came from. He compared their history to that of ancient people in South America and found similarities which he explored in *Kon-Tiki*.

Living in the culture you are studying can present difficulties. The first is language. An anthropologist must be able to understand a people's language to understand what is happening.

An anthropologist wants to see day-to-day life of a culture. But the anthropologist's presence may change the way people act. The people may show off for the anthropologist, or hide things that they think the anthropologist may not like or understand.

A third challenge can be the culture itself. For example, some cultures do not allow strange men near the women. A male anthropologist would not be able to find out how the women of the culture lived.

A fourth challenge is the anthropologist's own point of view, or bias. For example, a sharp pointed stick may look like a spear to an anthropologist. But it may really be used as a digging tool. An anthropologist must not assume that the culture he or she is studying is just like his or her own.

Anthropologists have shown us that we have a lot to learn from any culture. Anthropology has helped us understand our past and ourselves.

About Thor Heyerdahl

Thor Heyerdahl has become famous for testing his theories about how ancient people moved from place to place—by trying to do it himself.

Thor Heyerdahl was born in 1914 and grew up in Narvik, Norway. Heyerdahl attended the University of Oslo, in the capital city of Norway. He studied geography, Polynesian ethnology and zoology.

In 1936, Heyerdahl and his wife lived on Fatu Hiva, one of the islands in French Polynesia, for a year. Heyerdahl noticed that many of the plants on this tiny island in the middle of the ocean were the same as those in western South America, more than 4,000 miles to the east across the Pacific Ocean. How did they get there? He also noticed that the islanders' giant statues of their ancestor Tiki looked like statues made by primitive people of South America. The stories of Tiki were like those of an early king, Kon-Tiki, in South America.

When Heyerdahl went back to Norway, he read all the books he could on the people of the Pacific. Heyerdahl developed a theory that ancient people had come from the east on rafts and settled in Polynesia, bringing their plants and the story of their people with them.

During World War II, Heyerdahl served in the Norwegian armed forces in Europe. When the war was over, Heyerdahl was determined to prove that his theory might be correct. As the book *Kon-Tiki* describes, Heyerdahl made the voyage across the Pacific from South America to Polynesia in 1947 to prove that people *could* cross the vast ocean on a raft. The book *Kon-Tiki* was published in 1950 and became a bestseller.

After the book was published, many anthropologists and other scientists criticized Heyerdahl. They believed that the early people in the Pacific had come from Southeast Asia, not South America.

In 1958, Heyerdahl published *Aku-Aku*, the result of his research on Easter Island, a remote Polynesian island. Easter Island has very old, giant stone statues similar to those found in South America. Heyerdahl tried to prove that people who built the statues traveled from the coast of South America across the Pacific and settled on Easter Island.

In 1961, a conference of scientists who specialized in the study of the people and places in the Pacific agreed that the peoples of the Pacific islands came from *both* Southeast Asia and South America.

In 1969, Heyerdahl began another expedition, this time across the Atlantic Ocean from

Africa to the West Indies. He believed that thousands of years ago, the Egyptians could have traveled westward over the ocean to settle on the east coast of Central and South America.

He and a crew of six left the coast of West Africa and sailed westward on a 50-foot boat made of papyrus reeds. The boat was named *Ra* after the ancient Egyptian sun god. It sailed 2,700 miles across the Atlantic before it sank.

In 1970, Heyerdahl built *Ra II*, and this time the voyage was successful. He and the crew of eight reached the island of Barbados. He wrote *The Ra Expeditions* (1971) about the voyage.

In 1977, Heyerdahl set out in another reed boat to prove that the Sumerians, an ancient people who live in the Middle East, could have traveled in a boat from their country and settled in southwest Asia and the Arabian peninsula. The 4,000-mile trip from the Tigris River in Iraq to the Red Sea was successful. Heyerdahl wrote about the voyage in *The Tigris Expedition* (1980).

At age 78, Heyerdahl is back in Peru where he started the *Kon-Tiki* expedition 45 years earlier. He is working with other scientists to uncover the ruins of an ancient civilization.

Questions and Activities for the Reader

1. If you could talk to Thor Heyerdahl, what questions would you ask about his work? Think about how else you could find answers to these questions and talk about them with others.

2. Anthropologists and people who write about anthropology use special words to describe the things they study and write about. Because Thor Heyerdahl tried to prove his anthropological theory by sailing on a raft, this book has many words used in sailing. Some of these words are ones that you hear every day, like "surf," and others are more specialized and not very common, like "fathoms." You can find many of these specialized words in the Glossary on page 50. Do you use words in your work or hobby that are unique? Would others understand these words if you didn't explain them? Make a list of "special" words you use at work or in your other pursuits. How does your list compare with those of others?

3. Did reading the selections give you any

ideas for your own writing? You might want to write about:

• something you believe is true. What arguments would convince others?
• your powers of observation. What have you seen that others may have missed?
• people who live in different places but have similar customs. Why do you think these people have these customs in common?

4. Some archaeologists study materials left by ancient cultures. It is not always clear how an ancient object was used. Objects found in ruins are often broken. Imagine you are an archaeologist studying the remains of our present culture. Select a common household item. Imagine that you found the object in a ruin.

Describe the object as carefully as you can. Write only what the object looks like; do not say what it is used for.

Now read your description and see if you can guess what the object might be used for. For example, you might have described a potato peeler. But an archaeologist from the future might not know that our culture peels vegetables sometimes. You might instead guess that the tool you described, a pointed piece of metal with a handle and a sharp open groove, was a tool used for digging holes to plant seeds or a leatherworking tool. Write down what you think your object *could* be used for.

You might want to read your written description to a friend or family member. See if they can figure out from your description what the object really is.

5. Our culture is just as interesting and unusual as any ancient or far-off culture.

Take a notebook, a watch and a pencil or pen to a place where you can find a lot of people. Find a spot to begin your observations. What are people doing? Possible behaviors you will see might include: talking, window-shopping, buying items, eating and walking. Pick one kind of behavior to observe and study it for 15 minutes.

How many people did the behavior during your observation? What was the size of each group? How long did each person or group do the behavior? It might be easier if you create a tally sheet like the following one to help you keep track of your data.

WINDOW-SHOPPING BEHAVIOR

Number of Men	Number of Women	Number of Children	Total Group Size	Time Spent
1.				
2.				
3.				
4.				

After you have completed your study, look at your data. Try and find out information from the numbers. For example, what was the most common size of the groups that were talking? How long did most people take to eat food? Did more men window-shop than women? Can you form a theory based on your observations?

6. Imagine that you are the leader of an expedition. List all the supplies you would need for a long trip. Here are some things to think about while you are making your list:

a) Heyerdahl thought six people were the right number for his trip. How many would you take? Who would they be? What would their jobs be?

b) What kinds of food will you bring? How much?

c) What else do you need to live? Food and water are the most important things. Are there other things you will need such as first aid supplies or special tools and equipment?

d) What about entertainment? A long trip can be boring. What would you bring for fun?

RESOURCES

Books by Thor Heyerdahl

Kon-Tiki: Across the Pacific by Raft (1950).

American Indians in the Pacific: The Theory Behind the Kon-Tiki Expedition (1952).

Aku-Aku: The Secret of Easter Island (1958).

The Ra Expeditions (1971).

Fatu-Hiva: Back to Nature (1974).

The Tigris Expedition: In Search of Our Beginnings (1980).

Early Man and the Ocean: A Search for the Beginnings of Navigation and Seaborne Civilization (1981).

The Maldive Mystery (1986).

Kon-Tiki Man: An Illustrated Biography of Thor Heyerdahl, with Christopher Ralling (1991).

Magazines

National Geographic
Natural History
Smithsonian

Videos

Kon-Tiki, 1951
The Ascent of Man, 13 vol., 1974
Among the Wild Chimpanzees, 1984
Margaret Mead and Samoa, 1988

Four series of good books for all readers:

Writers' Voices—A multicultural, whole-language series of books offering selections from some of America's finest writers, along with background information, maps, glossaries, questions and activities and many more supplementary materials. Our list of authors includes: Amy Tan * Alex Haley * Alice Walker * Rudolfo Anaya * Louise Erdrich * Oscar Hijuelos * Maxine Hong Kingston * Gloria Naylor * Anne Tyler * Tom Wolfe * Mario Puzo * Avery Corman * Judith Krantz * Larry McMurtry * Mary Higgins Clark * Stephen King * Peter Benchley * Ray Bradbury * Sidney Sheldon * Maya Angelou * Jane Goodall * Mark Mathabane * Loretta Lynn * Katherine Jackson * Carol Burnett * Kareem Abdul-Jabbar * Ted Williams * Ahmad Rashad * Abigail Van Buren * Priscilla Presley * Paul Monette * Robert Fulghum * Bill Cosby * Lucille Clifton * Robert Bly * Robert Frost * Nikki Giovanni * Langston Hughes * Joy Harjo * Edna St. Vincent Millay * William Carlos Williams * Terrence McNally * Jules Feiffer * Alfred Uhry * Horton Foote * Marsha Norman * Lynne Alvarez * Lonne Elder III * ntozake shange * Neil Simon * August Wilson * Harvey Fierstein * Beth Henley * David Mamet * Arthur Miller and Spike Lee.

New Writers' Voices—A series of anthologies and individual narratives by talented new writers. Stories, poems and true-life experiences written by adult learners cover such topics as health, home and family, love, work, facing challenges, being in prison and remembering life in native countries. Many *New Writers' Voices* books contain photographs or illustrations.

Reference—A reference library for adult new readers and writers.

OurWorld—A series offering selections from works by well-known science writers, including David Attenborough, Thor Heyerdahl and Carl Sagan. Books include photographs, illustrations, related articles.

Write for our free complete catalog: Readers House/LVNYC, 121 Avenue of the Americas, New York, NY 10013